Short Reports from the International Science Meeting
come2graz - International Week 2013

University of Teacher Education Styria

Angela Gastager (Publisher)
Heiko Vogl (Publisher)

Copyright © 2013 Angela Gastager, Heiko Vogl

All rights reserved. No part of this book may be reproduced or transmitted by any person or entity, including internet search engines or retailers, in any form or by any means, electronic or mechanical, including photocopying, recording, scanning or by any information storage and retrieval system without the prior written permission of the publishers.

Angela Gastager, Heiko Vogl
Pädagogische Hochschule Steiermark
Hasnerplatz 12
8010 Graz
Austria

Graz, 2013

ISBN 978-1-291-43649-5 (pbk)
ISBN 978-1-291-43656-3 (ebk)

Table of Contents

Preface   5

Supporting New Teachers at the Beginning of their Professional Careers: A Collegial Mentoring Project in Styria, Austria
Andrea Holzinger, Karin da Rocha   7

Using Social Software on Erasmus Exchange: A Qualitative Study of How Erasmus Outgoing Students from the University College of Teacher Education are using Social Software during their Erasmus Student Exchanges
Heiko Vogl   9

Happiness Makes School
Eva-Maria Chibici-Revneanu, Maria Koppelhuber   13

An Analysis of Preschool Children's Perspectives on their Relationships with Adults
Angela Gastager, Bärbel Hausberger   15

Support of Plurilingualism of Students in Narva College of the University of Tartu, Estonia by Implementing CLIL and Trilingual Educational Model
Katri Raik, Nina Raud   19

Learning Space Deutschlandsberg – Creating Language Friendly Learning Environments
Martina Huber-Kriegler, Dagmar Gilly, Eva Theißl, Sonja Vucsina   23

MALEDIVE - Diversity in Majority Language Learning - Supporting Teacher Education
Dagmar Gilly   27

When Perception is Upside Down
Gertrude Jaritz, Birgit Schloffer   29

Inquiry-based Learning and Teaching for the Development of Students' Competences in Science
Eduard Schittelkopf, Veronika Rechberger, Erich Reichel 31

Living with Type 1 Diabetes
Andrea Lukács 33

Preface

In the context of the COME2Graz – International Week a Scientific Meeting was held on the 16th of May 2013 at the University College of Teacher Education Styria. Guests from different European Educational Institutions and eight research teams of our University College, in all about 30 persons including some international students, participated and talked about research in schooling and teaching. They shared their diverse experiences. Ten different research studies were presented, which concern five different fields of educational research. They run as follows.

Field 1 - Development of schools and professionalism

Karin da Rocha and Andrea Holzinger present their research about supporting new teachers at the beginning of their professional careers with a collegial mentoring project in Styria. Heiko Vogl explains how to use software on Erasmus exchange and reports a qualitative study how Erasmus outgoing students from the University College of Teacher Education are using social software during their Erasmus student exchanges. Eva-Maria Chibici-Revneanu and Maria Koppelhuber are engaged as researchers with Happiness in schooling.

Field 2 - Early childhood

Angela Gastager and Bärbel Hausberger analyze the perspective of preschool children on their relationship with adults.

Field 3 - Language

Katri Raik and Nina Raud from Estonia explain a study about the implementation of CLIL orientated teaching in the curricula. Martina Huber-Kriegler, Dagmar Gilly, Eva Theißl, and Sonja Vuscina report about how to create language friendly learning environments. Dagmar Gilly presents a European project about diversity in majority language learning. This project belongs also to the following field.

Field 4 - Diversity

Gertrude Jaritz and Birgit Schloffer speak about the special topic when the perception is upside down; it implies cerebral visually impaired children.

Field 5 - Natural Science and Health

Eduard Schittelkopf, Veronika Rechberger, and Erich Reichel demonstrate very vivid how functions inquiry-based learning and teaching for the development of students' competence in science. Andrea Lukacs from Hungary reports her study about how is live with diabetes type 1.

With pleasure we express many thanks to Dr. Barbara Pflanzl, head of the Institute 1 Research, Transfer of Knowledge and Innovation and to Mag. Susanne Linhofer, head of the Center 2 International Relations, for supporting us; and finally we say many thanks for the science lectorate to Kim Preston.

Angela Gastager
Heiko Vogl
Graz, 29.05.2013

Supporting New Teachers at the Beginning of their Professional Careers: A Collegial Mentoring Project in Styria, Austria

Andrea Holzinger, Karin da Rocha

Until now there was no possibility for new primary and secondary school teachers in Austria to gradually master the challenges at the beginning of their professional careers. They were left alone with full responsibility and, in most cases, a fulltime employment right from the beginning with only the practical experience they had gained during their three-year teacher training.

As about 900 primary and 1400 secondary school teachers are going to retire in Styria, Austria, in the next five years, the University of Teacher Education in Graz is carrying out a pilot project to aid new teachers in three Styrian regions accompanied by evaluation research from 2012 to 2014. Primary school teachers who are new to the job are advised by experienced colleagues at their local schools. Additionally, courses for mentors, mentees, and principals are offered. The lectures' design allows for social and digital networking by means of several meeting or online facilities. The contents of the courses are adapted to the roles and interests of the three focus groups and hold the possibility of peer discussion and cross-group communication. Some of the topics are as follows:

- Challenges of professional development for new teachers and mentoring basics (for principals).
- Classroom management, work with parents, and assessment of performance (for newcomers).
- Mentoring and feedback (for mentors).

The research accompanying the project investigates factors at structural, systemic, and personal levels that help new teachers to have a successful start in their chosen profession. As well, the value of collegial mentoring as a part of the human resources and school development is examined. Accordingly, the following research questions are pursued:

- Which factors can help to establish mentoring at school?
- How can mentoring help build professionalism in new teachers?
- How does specific training facilitate new teachers' starts in professional life?
- Which roles do social and digital networks play in the challenges of new teachers' professionalism?

During the two-year project, participants are asked to contribute to evaluation research by completing two questionnaires concerning mentoring at the beginning and at the end of the school year. Moreover, expert interviews are conducted three times a year.

First analyses have proved the necessity of the project, especially from the mentees and mentors' points of view, and have allowed for detailed insight into the fields new teachers are challenged with at the beginning of their careers like grading, administration, and meeting parents.

References:

Hericks, U. (2006). Professionalisierung als Entwicklungsaufgabe. Rekonstruktionen zur Berufseingangsphase von Lehrerinnen und Lehrern. Wiesbaden: VS Verlag für Sozialwissenschaften

Keller-Schneider, M. (2010). Entwicklungsaufgaben im Berufseinstieg von Lehrpersonen. Münster, New York, München, Berlin: Waxmann

Stöger, H., Ziegler, A., Schimke, D. (Hrsg.) (2009). Mentoring: Theoretische Hintergründe, empirische Befunde und praktische Anwendungen. Lengerich: Pabst

Contact: andrea.holzinger@phst.at; karin.darocha@phst.at

Using Social Software on Erasmus Exchange: A Qualitative Study of How Erasmus Outgoing Students from the University College of Teacher Education are using Social Software during their Erasmus Student Exchanges
Heiko Vogl

This study's focus is on the following question: How do Erasmus students from the University College of Teacher Education Styria use the internet and social software before, during and after their student exchange? This primary question can be divided into two aspects:

- Which social software applications do Erasmus students use?
- How do the software usages change during their exchange?

The study was framed by the model of social capital (Bourdieu, 1983). The term social capital is used in the literature for several different concepts. Depending on the different social science disciplines such as political science, economics, and sociology, the term is used in different ways. In sociology, social capital is referred to as resources, which the player do not have themselves, but achieve as the result of a network or on the basis of relations (Bourdieu, 1983). Social capital can be seen as the value of social relationships. The prerequisite is that this network of relationships must be established first and maintained afterwards. Nan Lin (1999) highlights the advantage of such relationships. Investments by individuals to such a network support the entire network. Furthermore, the individuals expect benefits and profits from such investments. Robert Putnam (1995) describes social capital through its different characteristics on the macro-social level. He speaks of networks, norms and trust from which individuals and the society itself benefit. Mutual benefit of social capital is in the center of his analysis. Network theorists like Lin, in contrast to Putnam, state that social capital is not located on the macro-social level. She sees social capital as a resource that is embedded in social structures. The use of such social structures according to Lin includes three elements: embedding, accessibility and usability. Following Bourdieu, Franzen and Pointner highlight the network-based dimension of social capital.

The different forms of social capital are similarly differentiated in scientific literature. For this study, strong/weak ties and bridging/bonding social capital (Putnam, 1995) were relevant.

The transfer of information between people with strong ties is smaller than between people with weak ties. There is a different level of information between people with weak ties. Individuals with weak ties help to bridge the information gap between different social communities ("bridging") by using social media. Putnam developed this model based on Granovetter's idea of bridging and binding forms of social capital.

Social software in general and online social networks like Facebook in particular enables the maintenance of weak and strong ties and the development of social capital. With different tools in different communities, relationships can be formed and maintained. The ways of supporting each other has changed through social software and social online networks. Social software makes it possible to support relations without being dependent on time and place.

This qualitative study, based on the "Grounded Theory", analyzes the individual user behavior of 6 outgoing Erasmus students in the academic year 2009/2010. The sample was compiled by analyzing student E-Portfolios and student Information and Communication Technologies (ICT)-competences and by balancing gender and geographical distribution. The students were interviewed between three and five months after their return. The interviews were recorded and guided with an interview manual developed using the SPSS (collecting, testing, sorting, subsume) method. Analyzing and coding the transcripts started after the first interview. The result of analyzing all data corresponds to a theoretical four-phase model.

A theoretical four-phase model was developed to show the students' use of the internet and of social software while staying abroad. In phase 0 (pre-mobile phase), they use the internet to collect information and to prepare for their stay abroad. In phase 1 (mobile phase 1), they use it to maintain social relationships with their home country. In phase 2 (mobile phase 2), they form new social relationships in the host country and join the local Erasmus community (peer group). The post-mobile phase (phase 3) is used for developing and maintaining social relationships, or even ending them, within the Erasmus community after returning home.

References:

Bourdieu, P. (1983). Ökonomisches Kapital, kulturelles Kapital, soziales Kapital. In R. Kreckel (Ed.), Kreckel (Trans.), *Soziale Ungleichheiten* (Vol. Sonderband 2). Retrieved from http://unirot.blogsport.de/images/bourdieukapital.pdf

Lin, N. (1999). Building a Network Theory of Social Capital'. Connections, (22), 28–51.

Putnam, R. D. (1995). Bowling Alone: America's Declining Social Capital. Journal of Democracy, 65–78.

Vogl, H. (2011). *Mit Facebook und Co. auf Erasmus: Eine sozialwissenschaftliche Studie über die Nutzung von Social-Software während des Erasmus-Studienaufenthaltes.* Grin Verlag.

Contact: heiko.vogl@phst.at

Happiness Makes School
Eva-Maria Chibici-Revneanu, Maria Koppelhuber

Teaching happiness as a school subject demonstrates that happiness in the sense of self-contentment and well-being can be trained and that personal responsibility may contribute to one's sense of happiness. The subject was first taught in the 2009 summer semester in 2 elementary schools, 2 comprehensive schools, 1 grammar school, and 1 institution of higher education. As of 2010, all Styrian schools were eligible to apply for the project. Subsequently, the number of "happiness" schools has grown to 96.

In the 2009 winter semester, a research project at the University of Teacher Education Styria was initiated to support the development of content and the implementation of the school subject in 6 Styrian pilot schools as well as the study of its effects on students, parents, and teachers. The research question focused on whether self-confidence and self-development of students can be improved by introducing and teaching this subject. Consequently, inhibitive as well as helpful criteria and measures, which were taken in order to implement happiness as a school subject, were analyzed. The research team also dealt with the topic of useful support structures that would be needed to install and develop the subject at schools. Mixed methods (case studies, quantitative as well as qualitative questionnaires, interviews with teachers and parents, and subjective observations) were applied.

In summary, the results showed that the school subject happiness has positive effects on students between the ages of 6 and 18:

- The atmosphere in the classrooms improved, and a rise in confidence (in students as well as in teachers) was observed.
- Students were more ready to bear responsibility.
- Teachers felt they responded more to their students' needs.
- Social competences were boosted.

- Students were more aware as well as proud of their personal achievements.

Contact: eva-maria.chibici-revneanu@lsr-stmk.gv.at

An Analysis of Preschool Children's Perspectives on their Relationships with Adults
Angela Gastager, Bärbel Hausberger

This study's main interest is the following question: Is the relationship between young children and adults one of respect and appreciation? The focus is on three aspects:

- Young children's competence as partners in research generally and their ability to report on matters that affect them. This study sought to pursue research conversations with children about their lived experiences in order to develop a richer understanding of children and childhood.

- Identifying ways in which children express their own experiences of daily communications with related persons. Children form their views and theories and give a clear indication of what constitutes good quality in this domain.

- Attitudes and opinions of preschool children are reconstructed, and analyzed along the lines stated in Article 12 of the U. N. Convention on the Rights of the Child (see http://www.hrweb.org/legal/child.html):

    1. States Parties shall assure to the child who is capable of forming his or her own views the right to express those views freely in all matters affecting the child, the views of the child being given due weight in accordance with the age and maturity of the child. 2. For this purpose, the child shall in particular be provided the opportunity to be heard in

> *any judicial and administrative proceedings affecting the child, either directly, or through a representative or an appropriate body, in a manner consistent with the procedural rules of national law.*

The study involved 26 young children in 2 preschool classes in Styria - 12 children in one class and 14 in the other. A preliminary warming-up phase allowed the participants to get familiar with one another. For the primary investigation, the so-called mosaic approach (Harcourt 2011) was used, which consists of whole group discussions, small group discussions, and finally individual or paired interactions. Children's drawings and conversations were used to consider and record children's views on and opinions of the quality of their own experiences, especially with the adults in their families and adults they meet in everyday life, such as streetcar drivers, dentists, pharmacists, waiters, etc. In particular, the children focus on their relationships within their own family, such as with their mother, father, grandparents, and siblings. The results show that children's ways of expressing their inner world and the corresponding articulation of their ideas vary extremely. For some of them, expressing and explaining their experiences is a great challenge. The children's class teacher (who is well-known to them) was very helpful in leading discussions and conversations with the children on the various topics, such as doing sports together, having celebrations at Christmas, and verbalizing the impressions that adult strangers leave in children's everyday lives. Two female preschool teachers were also involved in the research process. The subjective theories according to Gastager, Patry, and Gollackner (2011) were reconstructed in the dialogue-consensus-technique using the matrix analyzing technique developed in Salzburg by Patry and Gastager (2002) with a strong focus on innovative methodology. Results show that - according to one teacher - enhancing and blocking correlations between the main concepts of the cognition networks in the subjective theories are balanced, while - according to the other teacher - the correlations seem to be considerably unbalanced. However, the two teachers agree that the correlations within the concepts, enhancing and blocking appreciation and respect in the conversation between young children and adults, are very strong and very clear. Here are a few examples of important concepts that promise successful interaction for the persons concerned with this study: (1) a positive view of and attitude to children and to work in general; (2) a readiness to take one's time for a conversation; (3) children's self-confidence; (4) the teacher's condition and mood of the day; (5) politeness of interaction and manners. All in all, it may be concluded that respect and appreciation in the culture of interaction between young children and adults vary greatly and depends on the individuality of the child and the influences of situational conditions.

References:

Gastager, A. (2013). Subjektive Theorien von Lehrerinnen und Lehrern zum Umgang mit Vielfalt im Unterricht. *Erziehung und Unterricht*, 1-2, S. 108 - 117.

Gastager, A., Patry, J.-L., & Gollackner, K. (2011). Subjektive Theorien über das eigene Tun in sozialen Handlungsfeldern. Innsbruck: Studienverlag.

Harcourt, D. (2009). Standpoints on quality: Young children as competent research participants. Australian research Alliance for Children & Youth: NSW.

Harcourt, D. (2011). An encounter with children: Seeking meaning and understanding about childhood. European Early Childhood Education Research Journal, Volume 19, Issue 3, pages 331-343.

Contact: angela.gastager@phst.at; baerbel.hausberger@phst.at

Support of Plurilingualism of Students in Narva College of the University of Tartu, Estonia by Implementing CLIL and Trilingual Educational Model

Katri Raik, Nina Raud

Narva College of the University of Tartu, being located on the border of the EU with Russia, plays an important role in the higher education area of Estonia. Historically and geographically Narva is the place for coexistence of various languages and cultures. With the majority of the population of the region being Russian native speakers and with the Estonian language being the only state language of the country, Narva College offers higher education programmes in three languages (Estonian, Russian and English) and conducts researches into various aspects of multilingual and multicultural education at all educational levels (from kindergartens to universities). Its curricula follow the requirements of growing internationalization of general and tertiary education, which is commonly seen as one of the main aims of contemporary Europe in the current social, political, and economic context.

In view of the state requirements set to all university graduates in Estonia - the Estonian language mastery has to be at levels C1/B2 depending on students' area of specialization, and foreign languages mastery has to correspond to the same requirements of B2/C1 depending on the specialty – it is a College's priority to offer language modules that are aimed at developing students' language mastery. The volume of subjects varies with the focus being placed on the mastery of Estonian in the first place. Second foreign languages such as English, German and French are presented by general language studies and language courses for special purposes. However, Content and Language Integrated Learning (CLIL) plays a more important role while acquiring mastery in Estonian.

To provide qualitative higher education in three languages Narva College pays special attention to students' support and counselling to reveal problematic issues and assist students with appropriate help. As the language policy of the College is focused on developing language competences in three languages through teaching English/Estonian/Russian for special purposes and practicing content and language integrated teaching, one of the main College's concerns is how content studies in a foreign language are supported with appropriate teaching methods and language scaffolding; students' satisfaction with their academic studies and with their learning outcomes is also a key concern for both the administration and the academic staff of the College.

The report covers main aspects of the research into realization of CLIL practices in Narva College by presenting outcomes of the research conducted in Narva College with the support of European structural funds in 2011. The research was aimed at the analysis of class strategies employed by the college lecturers and on how principles of content and language integrated learning are used to create a safe and supportive educational environment for students. On the basis of the research outcomes a set of recommendations for both the administration and the academic staff was devised and suggested to improve quality of teaching in second/foreign languages in Narva College of the University of Tartu.

References:

Raik, K., Raud, N. 2012. Trilingual Educational Model in Narva College of the University of Tartu, Estonia: Issues and Challenges of the Implementation of CLIL in its Curricula. Available at http://www.sli-konferenz-mehrsprachigkeit.de/fileadmin/abstracts/sektion-5/Abstract_Raik_Raud.pdf Accessed on 16.06.2013

Burdakova, O., Dzalalova, A., Raud, N, 2011. CLIL Methodology in Narva Colelge of the University of Tartu (In Russian language) Available at http://www.narva.ut.ee/sites/default/files/narva_files/NK%20LAK%20uuringu%20raport%20vene%20k%202011%20_2_.pdf Accessed on 16.05.2013.

Contact: Katri.Raik@ut.ee; Niina.Raud@ut.ee

Learning Space Deutschlandsberg – Creating Language Friendly Learning Environments

Martina Huber-Kriegler, Dagmar Gilly, Eva Theißl, Sonja Vucsina

Deutschlandsberg is a mostly rural county southwest of Graz with several small villages and cities. This political unit has undertaken the task to transform all of its former Secondary Modern Schools into New Middle Schools at the same time while – in contrast to other counties – going through an accompanying coherent, long-term school development process. The project described here is a transfer project of a school development process initiated and developed by the Ministry of Education (2006-2011) called net1.

The two-year research project involves 9 NMS (the school development project will also involve 14 grade/primary schools in total), where two of the team members, Eva Theissl and Sonja Vucsina, work with varying faculty members over the course of all together three years. The overall goal of the school development process is to establish a cooperative and collaborative school climate in which individualized forms of learning can take place. A special role is attributed to space as an important constituent of the learning environment (the „Third Pedagogue"). The researchers focus on spatial arrangements and their implicit „messages" about the pedagogical approaches behind them, which they make visible through the method of „mapping".

In the course of the research project (accompanying the school development process), hundreds of photos were taken to document the areas of communication allocated to the students – either officially or informally „conquered" by the students. These photos were in the next step shown to the faculties of the schools and interpreted by the researchers and the faculties themselves. The predominant conclusion was that most schools lack specific, welcoming areas for student communication. Several schools took action and created cozy corners in or outside classrooms where students now can get together during breaks and talk. Another focus was on documenting which spatial arrangements (e.g. of tables and chairs in a classroom) would indicate that more learner-centered forms of instruction would take place in a school – or whether the claims the schools had often made in their published school profiles (on their websites) were counteracted by the usual rows of tables and chairs facing the blackboard in most classrooms, clearly indicating a more teacher-centered approach.

Students of 5 schools were also asked to film their favorite places in and around school, and these films were exchanged and shown to the film teams and faculties of other schools. The faculty was asked to give feedback to the film teams of their own schools, the students to each other. The 4 schools in the control group were not involved in that process. An innovative aspect of this research is the use of „mapping" as a research method – i.e. laying photographs next to each other and describing and interpreting the differences as results of development processes.

The next step will involve focusing on the forms of written communication inside and outside the school buildings and the degree to which schools with plurilingual students reflect this diversity in their official communication with students and parents. Are important messages translated into various L1s? Is L1 support of any kind available? Are students encouraged to attend L1 instruction? Are any languages other than German visible/audible in and around the school? Is language and language use a topic of instruction? And is the role of the language of instruction consciously seen and acknowledged by the teachers? Apart from the available photo documents, a number of school visits by members of the research team supplemented by interviews should help to clarify the answers to these questions. Finally the researchers hope to compile a list of recommendations on how to create school policies that, on one hand, stress the importance of communicative processes and, on the other hand, clearly and consciously foster all kinds of language development (L1, L2, L3) for students and teachers.

References:

Heil, C. (2007). Kartierende Auseinandersetzung mit aktueller Kunst. Erfinden und Erforschen von Vermittlungssituationen. München 2007

Legwie, H. (2000). Feldforschung und teilnehmende Beobachtung, In: Flick, Uwe u.a.(Hg.): Qualitative Sozialforschung. Ein Handbuch. München: Rowohlt

Schrittesser, I., Fraundorfer, A., Krainz-Dürr, M. (2012). Innovative Learning Environments. Wien: facultas.wuv.

Stiller, J. (in Vorb.). Gegen das blinde Sehen – empirische Rezeptionsforschung im Unterrichtfach Kunst. Dortmunder Schriften zur Kunst, Band 4.

Contact: martina.huber-kriegler@phst.at

MALEDIVE - Diversity in Majority Language Learning - Supporting Teacher Education
Dagmar Gilly

In multilingual schools the range of learners' first languages is wide. This means that the language taught in the majority language (ML; "mother tongue") classroom is L1 for only a few learners. Indeed, all the learners in the classroom are plurilingual, as they learn many foreign languages in school and master different varieties of their first language. ML teaching has thus far been kept separate from other language subjects. However, in multilingual schools we should not ignore learners' proficiency in various languages; rather, all language teaching should enhance learners' individual and multilayered language repertoires and support the development of a holistic linguistic identity. This is essential in developing literacies and effective learning and teaching for all learners.

In the Maledive project (Diversity in Majority Language Learning - Supporting Teacher Education - http://maledive.ecml.at/) we aim to develop concrete tools and study modules for changing the mindset in ML teacher education from monolingual to plurilingual and promote collaboration between teachers of all languages and also other subjects. The developmental work is done in close collaboration with international networks in many workshops arranged in different countries. The project is part of a program operated by the European Centre for Modern Languages (ECML – http://ecml.at), the mission of which is to encourage excellence and innovation in language teaching and to help Europeans learn languages more efficiently.

When developing a more pluralistic approach to majority language teaching, the key issues are

- How can learners' plurilingual repertoire and intercultural competences be developed?

- How can a plurilingual approach be embedded in the ML curriculum and integrated with other learning contents?

- How can productive cooperation and shared vision between teachers of different types of languages (majority language, foreign languages, second languages, first languages) be enhanced?

- What kinds of approaches can be used for developing learners' language repertoire in the majority language classroom?

- How can teacher education be developed to prepare teachers to practice more inclusive, plurilingual approaches?

References:

Project Handbook:
http://marille.ecml.at/Handbook/tabid/2597/language/en-GB/Default.aspx

Belke, Gerlind (2012): Mehr Sprache(n) für alle: Sprachunterricht in einer vielsprachigen Gesellschaft. Baltmannsweiler: Schneider Hohengeren.

Busch, Brigitta (2013): Mehrsprachigkeit, *UTB*: facultas.wuv .

Cummins, J. & Early, M. (Eds) (2011) Identity texts: The collaborative creation of power in multilingual schools. Stoke-on-Trent: Trentham Books.

Contact: dagmar.gilly@phst.at

When Perception is Upside Down
Gertrude Jaritz, Birgit Schloffer

„Cerebral visual impairment is one of the most common eye diseases in the Western world" (Gordon Dutton).

Many visually impaired children in Austria have Cerebral Visual Impairment (CVI). Unfortunately, the assessment process is inefficient and there is little confidence in the results. This casts doubt on the development of learning strategies. The project team members began a research project within the framework of the University College of Teacher Education Styria and the Odilien Institute (Odilien-Institut Graz), a school for the visually impaired to answer the questions of how children with CVI can see and how better learning strategies can be created for these children. Prof. Lea Hyvärinen, an ophthalmologist from Finland (via TU Dortmund), and Dr. Marjolein Dik, a neuropsychologist from the Netherlands, graciously agreed to participate in the project as cooperative partners.

The initial phase of research took place in 2010-2011. When the project began, there was neither medical and neuropsychological diagnosis nor collaboration between the disciplines. Therefore, it was necessary to start with a visual assessment of the participating students. First, the families and the vision teachers provided their perspectives in a questionnaire adapted from Gordon Dutton's. The parents were also asked to provide all relevant medical reports. Then Prof. Hyvärinen examined the children and made a functional diagnosis in the context of vision and CVI after the ICF scheme (International Classification of Functioning Disability and Health) related to her four leaves of clover of visual functioning.

In the second phase of the project in 2011-2012, the neuropsychological assessment and IQ results were correlated with the vision test results and the completed questionnaires from the parents. The results were discussed with the parents and the teachers were subsequently informed about the conditions of the children and best learning possibilities.

The third and final phase has begun and will be completed in 2014. Meetings of teachers, parents, and children in special focus groups are on going to collect strategies and best learning examples. As we are still in the process of research, first outcomes can be presented in this Round Table of ISM.

Reference:

Dutton, G.N. & Bax, M. [Hrsg.] (2010). Visual Impairment in Children Due to Damage to the Brain. London.

Contact: gertrude.jaritz@phst.at

Inquiry-based Learning and Teaching for the Development of Students' Competences in Science

Eduard Schittelkopf, Veronika Rechberger, Erich Reichel

In Germany, Switzerland, and Austria a lot of work has been done in introducing models for supporting competences in physics, chemistry, and biology lessons. The Austrian model is a suitable tool for the preparation of modern science lessons and is divided into three dimensions: competences, level of performance, and content. The competences are split into three main parts: organization of knowledge (working with different sources), achievement of knowledge (contains mainly scientific methods, e.g. experiment), and drawing of conclusions.

In our work, we focus on the competence of achieving knowledge. Therefore, the observation of a particular phenomenon by the students plays an important role. By means of suitable experiments, students are encouraged to note their own observations. In the next step students write down questions they would like to have answered. After the formulation of their own research question they try to answer it by themselves - they develop their own hypothesis. This hypothesis will then be proved by a student-developed experiment, which is an improved or altered version of the starting experiment. The students' own experiment leads to further observations, to new questions, and so on. In this way a learning cycle is introduced that enables research and, thereby, inquiry-based learning in the classroom. It is not necessary to run through the whole cycle every lesson. Depending on the content of the lesson, only some phases of the cycle are focused on.

To help teachers introduce this cycle in their own lessons, we developed easy to use tools. For example, the students fill in a protocol sheet during their work that allows the teacher to count the number of observations. Then the quality of the observations according to the phenomenon is determined. A second tool allows the teacher to find out the most suitable experiment for the training of particular competences.

We used this method in various classes with different experiments and found it to be an applicable method for developing competences in physics lessons. The tools proved to be helpful in detecting the students' competence increase.

References:

Bifie (2011). Kompetenzmodell Naturwissenschaften 8. Schulstufe.
> Verfügbar unter
> https://www.bifie.at/system/files/dl/bist_nawi_kompetenzmodell-8_2011-10-21.pdf [19.3.2012]

Bonnstetter, R. J. (1998). Inquiry: Learning from the Past with an Eye on the Future. Electronic Journal of Science Education, 3 (1). Verfügbar unter http://www.scholarlyexchange.org/ojs/index.php/EJSE/article/view/7595/5362 [19.3.2012]

Reichel, E. & Schittelkopf, E. (2012). Förderung von experimentellen Kompetenzen - Forschendes Lernen in vier Phasen. IMST Newsletter 37.

Contact: erich.reichel@phst.at

Living with Type 1 Diabetes
Andrea Lukács

Type 1 diabetes (T1D) is an autoimmune disease that tends to occur in childhood, adolescence or early adulthood, but it may have its clinical onset at any age. There is no recovery from the disease and the patient needs life-long exogenous insulin. The incidence of T1D is rising in the young population. The annual increase in Hungary is 4.4% in the last two decades. Children and adolescents spend a lot of time at school without their parents who supervise the disease. Diabetes care is important in the school for the immediate and long-term well being, and even for the optimal academic performance. Teachers should know if a child has diabetes so they can provide help if needed. It is important to understand how the disease influences youths' health-related quality of life (HRQoL). The aim of this study was to evaluate the HRQoL of children and adolescents with T1D using child self-report (CSR) and parent proxy-report (PPR), and to compare their HRQoL to the healthy peers.

A total of 355 diabetic youths (184 boys and 171 girls) with their parents (n=328) and 294 randomly chosen healthy children and adolescents (137 boys and 157 girls) with their parents (n=294) took part in this cross-sectional survey. All the participants were between aged 8-18. HRQoL was assessed with the Pediatric Quality of Life Inventory (PedsQL) Generic Core Scale (GCS) and 3.0 Diabetes Module (DM).

Comparing the diabetic and the non-diabetic groups by gender on the basis of GCS we found no statistically significant differences in quality of life neither in CSR or PPR, except of the Physical functioning in boys by the PPR. The parents rated the physical functioning significantly better for control boys than diabetic boys (p=0.005). The children and the parents' concordance showed similarity in healthy groups. The parents of the diabetic group significantly underestimated their children' HRQoL in all subscales of the GCS (Physical functioning CSR: 82.50 ±10.90 vs. PPR: 78.83 ±10.23; p<0.001, Emotional functioning CSR: 71.30 ±16.73 vs. PPR: 66.97 ±16.36;

p<0.001, Social functioning CSR: 90.22 ±13.86 vs. PPR: 87.23 ±15.08; p<0.001, School functioning CSR: 74.27 ±14.78 vs. PPR: 72.67 ±14.76; p=0.003). It was the case in the DM. The parents significantly underestimated their children's HRQoL in all subscales except of Communication subscale (Diabetes symptoms: CSR: 64.57 ±13.27 vs. PPR: 62.60 ±12.30; p<0.001, Treatment barriers: CSR: 70.27 ±19.81 vs. PPR 65.47 ±20.03; p<0.001, Treatment adherence: CSR: 83.58 ±13.32 vs. PPR: 80.10 ±14.17; p<0.001, Worry: CSR 69.87 ±20.43 vs. PPR: 62.91 ±21.30; p<0.001, Communication: CSR: 78.30 ±22.17 vs. PPR: 76.93 ±22.39; p=0.123). Analyzing the DM subscale scores we found that diabetic youths had no problems with the treatment adherence and communication, but they had low scores in the diabetes symptoms and the worry subscales. A similar pattern was found in the PPR.

In conclusion, children and adolescents with T1D live similar lives to their healthy peers, but due to their chronic illness, they have special problems that they can cope with. The diabetes somatic symptoms have a negative effect on their quality of life and can occur at any place and at any time. Knowledge of these symptoms may be important for the teachers who could give more freedom to the child, even during the lesson, to manage the situation. Parents seem to overprotect their chronically ill children. They suffer the disease worse than the children do themselves. The long-term parental fear may limit the diabetic child's self-esteem and build panic issues in children as well, which may reduce both the child and parent efficacy for disease management. The role of the teacher can be important between the parent and child disagreement.

Reference:

Gyürüs E, Patterson CC, Soltész Gy, and the Hungarian Childhood Diabetes Epidemiology Group. (2012). Twenty-one year of prospective incidence of childhood type 1 diabetes in Hungary – the rising trend continues (or peaks and highlands?). Pediatric Diabetes, 13, pp. 21-25

Silverstein J, Klingensmith G, Copeland K, et al. (2005). Care of children and adolescents with type 1 diabetes: A statement of the American Diabetes Association. Diabetes Care, 28, pp.186-212.

Klingensmith G, Kaufman F, Schatz D, et al. (American Diabetes Association). (2004). Diabetes care in the school and day care setting. Diabetes Care, 27(suppl 1) pp. S122-S128.

American Diabetes Association. (2011). Diabetes Care in the School and Day Care Setting. Diabetes Care, 34, (Suppl 1) pp. S70-S74.

Contact: efklandi@uni-miskolc.hu

WRITTEN MENTION

Lifelong Learning Programme

This project has been funded with support from the European Commission. This publication [communication] reflects the views only of the author, and the Commission cannot be held responsible for any use which may be made of the information contained therein.

www.ingramcontent.com/pod-product-compliance
Lightning Source LLC
Chambersburg PA
CBHW072307170526
45158CB00003BA/1230